Herman likes the ocean.

But Herman squirms inside his shell.
His shell is not big enough.

"I need more space," said Herman.

"I guess so," answered Starfish.

"That shell has enough space," said Starfish.

Herman thanked his friend.

"I won't outgrow this," said Herman.

"Think twice!" said Starfish. "Young crabs need more space as they grow."